MAGIC SQUARE

Its New Discoveries
and Applications

DAVID MEI

ARPress

ILLUMINATING IDEAS,
EMPOWERING VOICES

ARPress
45 Dan Road Suite 36
Canton MA 02021

Hotline: 1(800) 220-7660
Fax: 1(855) 752-6001

Ordering Information:
Quantity Sales. Special discounts are available on quantity purchases by corporations, associations, and others. For details, contact the publisher at the address above.

Printed in the United States of America.

| ISBN-13 | Paperback | 979-8-89676-224-9 |
| | eBook | 979-8-89676-225-6 |

Library of Congress Control Number: 2024925141

The new discovery of construction of Magic Square enables us to:

1. Resolve the mysteries that have puzzled mathematicians for centuries.

2. Construct any size of magic with nothing but pen and paper.

3. Have some magic squares custom-made that means a desired number can be put in a desired cell.

4. Produce a large quantity of combinations for most magic squares in an easy and simple way.

5. Attribute the renewal of Magic Square Puzzle, the outstanding application of Magic Square.

6. Encourage youngsters to experience in learning many mathematical skills and in a sense of satisfaction & fulfillment while playing in a game setting environment that the traditionally mathematical establishment fails to accommodate.

CONTENTS

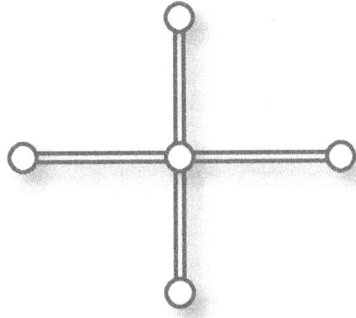

I

Magic Square Before Our Time

A magic square is merely a group of numbers arranged in a fancy way. Its existence dates back to ancient time. Philolaus once has said, "Number is the bond of the eternal continuance of thing." The mysterious relationships of numbers have occupied the minds of men in all ages through centuries since men incept numbers for simple counting to complicated calculations. The study of numbers becomes hit-parade those days and attributes the birth of magic squares. A square is called magic, only if the sum of numbers in each of its rows, columns, and diagonals adds up to a same amount.

One of the oldest magic square in the East in the Chinese Scroll of Loh attributed to Fuh-Hi, the mythical founder of Chinese civilization, who according to Chinese report lived 2858-2738 B.C.. The Scroll of Loh is a 3x3 magic square: all the odd numbers in whore dots are the symbol of Yang, the emblem of heaven, while the even numbers in black dots are the symbol of Tin, The emblem of the earth. Same as to their Eastern counterpart, the magic squares are only associated with astrologers, magicians, and talisman makers in the West, not until the mathematical value of magic square has been justified. From a paper entitled "Magic Squares and Other Problems on a Chessboard" by

Major P.A. Mac Mahon, R.A., D. Sc., F.R.S., published in Proceedings of the Royal Institute of Great Britain, in 1892, he concludes that the magic square intimately connected with the infinitesimal calculus and the calculus of finite differences.

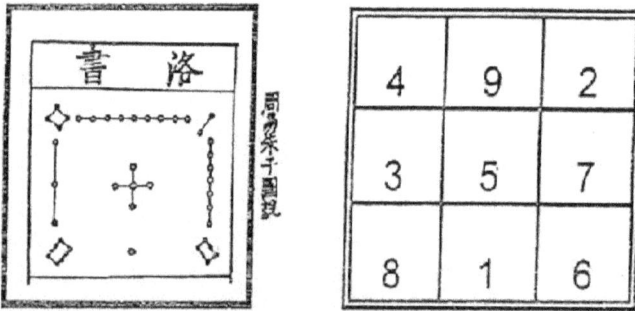

So, the study of magic square becomes the hottest subject for philosophers, mathematicians and the pass time for the noble class who can afford the time and luxury in Europe. However, they end up finding a very difficult time in constructing new magic squares. In late 19th century, W.S. Andrew, a famous mathematician from England, writes in his book Magic Square and Cubes that, "Many attempts have been made to construct magic squares from a natural series of numbers by locating each succeeding number a Knight's move from the last one, until every call in the square is included in one continuous Knight's tour. This difficult problem however has never been solved, and the square in question probably does not exist. Many squares have been made that sum correctly in their lines and columns, but they all fall in their two diagonals and therefore are not strictly magic." He also confirms that the construction of magic squares become laborious.

Nevertheless, the general public takes it as a golden opportunity for them to hit the jackpot. They compete each other by constructing magic square one by one, most of them bearing their names. Two of the following square are not even magic but all well known, the Jaina

Square, one of the oldest 8x8 square famed for its equilibrium and the Franklin Square of 8x8 and 16x15, created by Benjamin Franklin, L.L.L., F.R.S., one of our founding Father, famed for their astonished properties. There is another one by C.A Browne, Jr., a real 27x27 magic square with center number of 365, which represents an extreme significance of our calendar system referring to the days of the year and cycles of year.

THE JAINA SQUARE

10	5	16	3	10	5	16	3
15	4	9	6	15	4	9	6
1	14	7	12	1	14	7	12
8	11	2	13	8	11	2	13
10	5	16	3	10	5	16	3
15	4	9	6	15	4	9	6
1	14	7	12	1	14	7	12
8	11	2	13	8	11	2	13

THE FRANKLIN SQUARE 8 X 8

52	61	4	13	20	29	36	45
14	3	62	51	46	35	30	19
53	60	5	12	21	28	37	44
11	6	59	54	43	38	27	22
55	58	7	10	23	26	39	42
9	8	57	56	41	40	25	24
50	63	2	15	18	31	34	47
16	1	64	49	48	33	32	17

FRANKLIN SQUARE 16X16

C. A. BROWNE, JR. 27X27 MAGIC SQUARE

352	381	326	439	468	413	274	303	248	613	642	587	700	729	674	535	564	509	118	147	92	205	234	179	40	69	14
327	353	379	414	440	466	249	275	301	588	614	640	675	701	727	510	536	562	93	119	145	180	206	232	15	41	67
380	325	354	467	412	441	302	247	276	641	586	615	728	673	702	563	508	537	146	91	120	233	178	207	68	13	42
277	306	251	355	384	329	433	462	407	538	567	512	616	645	590	694	723	668	43	72	17	121	150	95	199	228	173
252	278	304	330	356	382	408	434	460	513	539	565	591	617	643	669	695	721	18	44	70	96	122	148	174	200	226
305	250	279	383	328	357	461	406	435	566	511	540	644	589	618	722	667	696	71	16	45	149	94	123	227	172	201
436	465	410	271	300	245	358	387	332	697	726	671	532	561	506	619	648	593	202	231	176	37	66	11	124	153	98
411	437	463	246	272	298	333	359	385	672	698	724	507	533	559	594	620	646	177	203	229	12	38	64	99	125	151
464	409	438	299	244	273	386	331	360	725	670	699	560	505	534	647	592	621	230	175	204	65	10	39	152	97	126
127	156	101	214	243	188	49	78	23	361	390	335	448	477	422	283	312	257	595	624	569	582	711	656	517	546	491
102	128	154	189	215	241	24	50	76	336	362	388	423	449	475	258	284	310	570	596	622	657	683	709	492	518	544
155	100	129	242	187	216	77	22	51	389	334	363	476	421	450	311	256	285	623	568	597	710	655	684	545	490	519
52	81	26	130	159	104	208	237	182	286	315	260	364	393	338	442	471	416	520	549	494	598	627	572	676	705	650
27	53	79	105	131	157	183	209	235	261	287	313	339	365	391	417	443	469	495	521	547	573	599	625	651	677	703
80	25	54	158	103	132	236	181	210	314	259	288	392	337	366	470	415	444	548	493	522	626	571	600	704	649	678
211	240	185	46	75	20	133	162	107	445	474	419	280	309	254	367	396	341	679	708	653	514	543	488	601	630	575
186	212	238	21	47	73	108	134	160	420	446	472	255	281	307	342	368	394	654	680	706	489	515	541	576	602	628
239	184	213	74	19	48	161	106	135	473	418	447	308	253	282	395	340	369	707	652	681	542	487	516	629	574	603
604	633	676	691	720	665	526	555	500	109	138	83	196	225	170	31	60	5	370	399	344	457	486	431	292	321	266
579	605	631	666	692	718	501	527	553	84	110	136	171	197	223	6	32	58	345	371	397	432	458	484	267	293	319
632	577	606	719	664	693	554	499	528	137	82	111	224	169	198	59	4	33	398	343	372	485	430	459	320	265	294
529	558	503	607	636	581	685	714	659	34	63	8	112	141	86	190	219	164	295	324	269	373	404	347	451	480	425
504	530	556	582	508	634	660	686	712	9	35	61	87	113	139	165	191	217	270	269	322	348	374	400	426	452	478
557	502	531	635	580	609	713	658	687	62	7	36	140	85	114	218	163	192	323	268	297	401	346	375	479	424	453
688	717	662	523	552	497	610	639	584	193	222	167	28	57	2	115	144	89	454	483	428	289	318	263	376	405	350
663	689	715	498	524	550	585	611	637	168	194	220	3	29	55	90	116	142	429	455	481	264	290	316	351	377	403
716	661	690	551	496	525	638	583	612	221	166	195	56	1	30	143	88	117	482	427	456	317	262	291	404	349	378

THE SUM OF 27 NUMBERS ON EACH ROW, COLUMN AND DIAGONAL IS CONSTANT 9855.

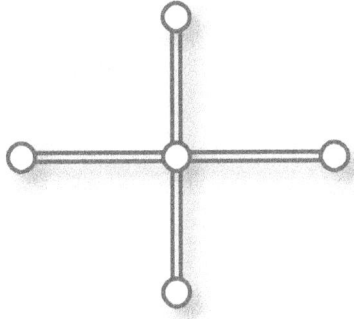

II

The Construction of A Magic Square

How to construct a magic square has always been an interesting topic for centuries. The following is going to show many ways to do it based on the size:

A) 3x3 Magic Square

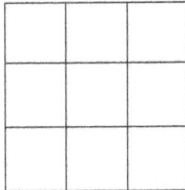

- We will start with a three by three (3x3), which is the simplest magic square you can make.

- It has nine empty boxes that you have to fill in the numbers starting from one to nine: 1,2,3,4,5,6,7,8,9

The following steps you have to take.

	5	

Step 1: Put the number 5 in the center

If you cross out the number 5 above, you have eight numbers left:

Odd number: 1, 3, 7, 9 and even numbers: 2, 4, 6, 8

2		4
	5	
6		8

Step 2: Put the four even numbers in the corners, but do it so the diagonals add up to 15. For example, 2+5+8=15 and 4+5=6+15

2	9	4
7	5	3
6	1	8

Step 3: Put the four odd numbers in the empty boxes, but do it so the horizontals and the verticals all add up to 15.

Horizontals: 2+9+4=15, 7+5+3=15, 6+1+8=15
Verticals: 2+7+6=15, 9+5+1=15, 4+3+8=15

The above 3x3 magic square can have 7 more by switching rows and/or columns. They belong to the same family. The only difference is their positions. It can be applied to all magic squares that call The Rule of Magic Square Family of Eight.

2	7	6
9	5	1
4	3	8

4	3	8
9	5	1
2	7	6

4	9	2
3	5	7
8	1	6

6	7	2
1	5	9
8	3	4

6	1	8
7	5	3
2	9	4

8	1	6
3	5	7
4	9	2

8	3	4
1	5	9
6	7	2

B) 4x4 Magic Square

Let us pick the easiest way to make a 4x4 magic square:

Method 1- a) Fill in the numbers from 1 to 16 as shown. **b)** Keep the four numbers in the center block and in each of the corner as they are, but switch the numbers in other spaces as shown. Then, you have an instant 4x4 magic square.

1	2	3	4
5	6	7	8
9	10	11	12
13	14	15	16

	2	3	
5			8
9			12
	14	15	

1	15	14	4
12	6	7	9
8	10	11	5
13	3	2	16

Method 2- a) Fill in the numbers from 1 to 16 as shown. **b)** Leave the numbers in the four corners unchanged but switch the numbers in two middle rows and columns as shown. **c)** Switch the numbers in the center block diagonally. Then, you have another 4x4 magic square.

1	2	3	4
5	6	7	8
9	10	11	12
13	14	15	16

	2	3	
5			8
9			12
	14	15	

6	7
10	11

1	14	15	4
8	11	10	5
12	7	6	9
13	2	3	16

1	15	14	4
12	6	7	9
8	10	11	5
13	3	2	16

Method 3- Let us try R/C Method, a newly discovered for smaller size of magic square. R represents ROW and C represents COLUMN. The idea is to switch rows and columns of an existing magic square. Additional magic squares can be produced as resulted from the validity of the two diagonals. Take the one 4x4 magic square on the left as the original for example. There will be 24 squares as resulted by switching 4 columns but only 8 of them come out magic as indicated in shade. Each of them creates another 24 4x4 squares. Among them, only 192 come out magic. Another 192 additional 4x4 magic squares can be obtained when the original 4x4 magic square changes its rows into columns.

4x4 (continued)

1	12	8	13
4	9	5	16
14	7	11	2
15	6	10	3

4	9	5	16
1	12	8	13
14	7	11	2
15	6	10	3

14	7	11	2
1	12	8	13
4	9	5	16
15	6	10	3

15	6	10	3
1	12	8	13
4	9	5	16
14	7	11	2

1	12	8	13
4	9	5	16
15	6	10	3
14	7	11	2

4	9	5	16
1	12	8	13
15	6	10	3
14	7	11	2

14	7	11	2
1	12	8	13
15	6	10	3
4	9	5	16

15	6	10	3
1	12	8	13
14	7	11	2
4	9	5	16

1	12	8	13
14	7	11	2
4	9	5	16
15	6	10	3

4	9	5	16
14	7	11	2
1	12	8	13
15	6	10	3

14	7	11	2
4	9	5	16
1	12	8	13
15	6	10	3

15	6	10	3
4	9	5	16
1	12	8	13
14	7	11	2

1	12	8	13
14	7	11	2
15	6	10	3
4	9	5	16

4	9	5	16
14	7	11	2
15	6	10	3
1	12	8	13

14	7	11	2
4	9	5	16
15	6	10	3
1	12	8	13

15	6	10	3
4	9	5	16
14	7	11	2
1	12	8	13

1	12	8	13
15	6	10	3
4	9	5	16
14	7	11	2

4	9	5	16
15	6	10	3
1	12	8	13
14	7	11	2

14	7	11	2
15	6	10	3
1	12	8	13
4	9	5	16

15	6	10	3
14	7	11	2
1	12	8	13
4	9	5	16

1	12	8	13
15	6	10	3
14	7	11	2
4	9	5	16

4	9	5	16
15	6	10	3
14	7	11	2
1	12	8	13

14	7	11	2
15	6	10	3
4	9	5	16
1	12	8	13

15	6	10	3
14	7	11	2
4	9	5	16
1	12	8	13

(COLUMNS ARE RE-ARRANGED IN ASCENDING ORDER FROM THE ORIGINAL)

Please note that the original square needs not to be magic, as long as each row and column adds up to 34.

C) 5x5 MAGIC SQUARE:

Method 1–The Diagonal Move Method (DMM)

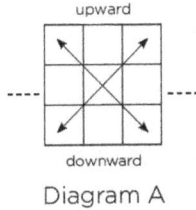

Diagram A

The DMM is one of two simple and easy ways to make odd number magic square for centuries. It is to put each succeeding number in a diagonal movement which has four directions as shown in Diagram A. Once you decide where to start and what direction to follow, you can proceed until you hit with two following situations:

1) OUT OF BOUND: Once a number goes out of the square, you have to bounce back the number into a corresponding position inside the square. See Diagrams B, C and D.

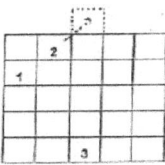
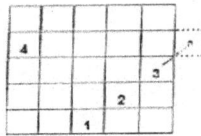
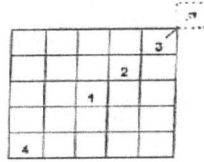

Diagram B   Diagram C   Diagram D

2) GETTING BLOCKED: When you bounce back the number and find the corresponding position occupied, you have to place that number directly under the proceeding number if it is an upward move or above if downward move. See Diagrams E and F.

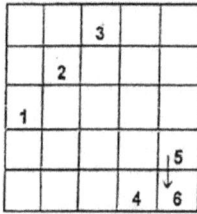

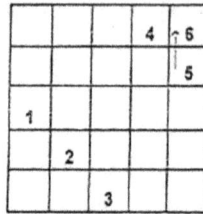

Diagram E   Diagram F

5x5 (Continued)

Not every cell or any direction can produce 5x5 magic square with DMM. In Diagram G, there are only 16 cells with 20 specific directions passing the test. In fact, DMM makes only 5 different 5x5 magic squares.

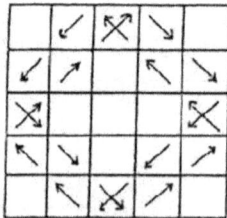

Diagram G

Method 2 – The Knight's Move Method (KMM)

The KMM is another simple and easy way to make odd number magic squares for centuries. It is so names because it uses the knight's movement from chess. There will be no restriction in using KMM for making 5x5 magic squares which means you can start any cell in any direction. However, you have to follow that in DMM, when the number you are going to put into the cell gets out of bound or blocked. There are 8 different knight's moves as shown in Diagram H, that produce 200 5x5 magic squares. Now you can compare the 2 5x5 magic squares constructed by above methods with same cell to start and direction.

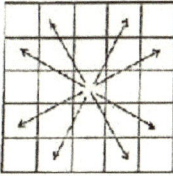

17	24	1	8	15
23	5	7	14	16
4	8	13	20	22
10	12	19	21	3
11	18	25	2	9

DMM

10	18	1	14	22
11	24	7	20	3
17	5	13	21	9
23	6	19	2	15
4	12	25	8	16

KMM

Diagram H

Method 3 – R/C Method

You have to be very patient in using the R/C Method to make 5x5 magic squares, because there are only 8 that come out magic in 120 switches of columns from the 5x5 magic square just made by DMM above.

D) 6x6 Magic Square:

Mei's Square Method (MSM) has been recently discovered for constructing even number magic squares. It is simple and easy to learn. It can create a large quantity of even number magic squares. It uses the idea that all even squares are of the form 4N and that each even square can be represented by a factor N and a 2x2 grid. The following steps are used for the making:

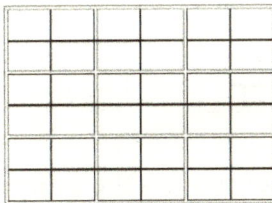

1) Divide the 6x6 square into groups of 2x2 square.

2) Label each of the 9 2x2 squares as if it were a 3x3 magic square (A 3x3 magic square is used as the BASE.)

13

8	3	4
1	5	9
6	7	2

3) Group the numbers 1 -16 into 9 groups of 4 so that they can be placed into each one of the 9 2x2 squared shown above.

Each group corresponds to a number in the 3x3 magic square of 2).

(1) 1,2,3,4   (2) 5,6,7,8   (3) 9,10,11,12

(4) 13,14,15,16   (5) 17,18,19,20   (6) 21,22,23,24

(7) 25,26,27,28   (8) 29,30,31,32   (9) 33,34,35,36

4) A Mei's Square for 6x6 is consisted of 9 patterns which show how 4 numbers situated. The open end tells where to begin and the bar tells where to end. See the following 3 samples below.

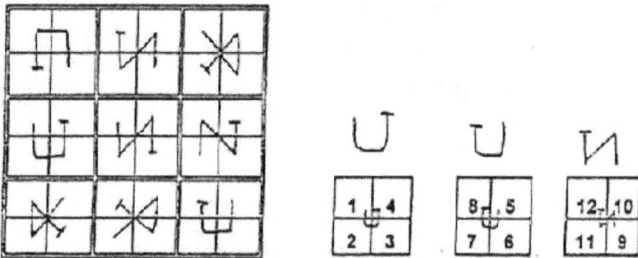

Now, all you have to is to connect the 3x3 magic square in 2) which shows the order of 9 numbers, and the Mei's Square for 6x6 which dictates 9 patterns of 4 numbers. Then, a 6x6 magic square has been created as follows.

31	30	12	10	13	15
32	29	11	9	16	14
1	4	17	19	34	36
2	3	18	20	33	35
23	21	28	26	8	5
22	24	25	27	7	6

E) 7x7 Magic Square

METHOD 1 – DMM

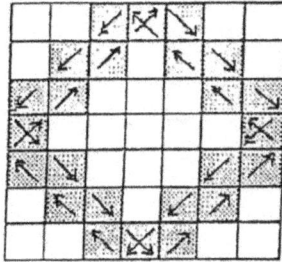

Apply the same operation of DMM in creating 5x5 magic square. Only this time, you have 24 cells and 28 directions to start with that produce 56 7x7 magic squares. See one on the left below.

METHOD 2 – KMM

33	42	44	4	13	15	24
41	43	3	12	21	23	32
49	2	11	20	22	31	40
1	10	19	28	30	39	48
9	18	27	29	38	47	7
17	26	35	37	46	6	8
25	34	36	45	5	14	16

33	43	11	28	38	6	16
41	2	19	29	46	14	24
49	10	27	37	5	15	32
1	18	35	45	13	23	40
9	26	36	4	21	31	48
17	34	44	12	22	39	7
25	42	3	20	30	47	8

Like 5x5, there is no restriction on making 7x7 magic squares with KMM which produces 1,568 7x7 magic squares. See one on right below.

Note: Please compare the two with both methods starting at same cell and direction.

F) 8x8 Magic Square

The MSM will be used to construct 8x8 magic square, so steps described in creating 6x6 magic square can be applied here. However, you need this time a 4x4 magic square as the BASE and a Mei's Square for 8x8 as shown below:

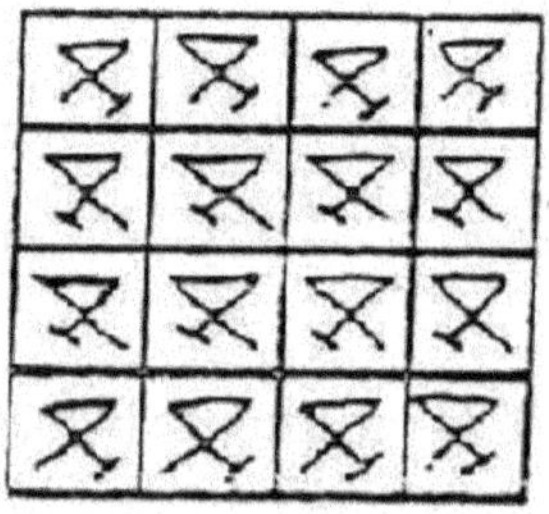

1	15	10	8
5	14	11	4
16	3	6	9
12	2	7	13

1) 1,2,3,4	2) 5,6,7,8	3) 9,10,11,12	4) 13,14,15,16
5) 17,18,19,20	6) 21,22,23,24	7) 25,26,27,28	8) 29,30,31,32
9) 33,34,35,36	10) 37,38,39,40	11) 41,42,43,44	12) 45,46,47,48
13) 49,50,51,52	14) 53,54,55,56	15) 57,58,59,60	16) 61,62 63,64.

Let us take the simplest way to group 64 numbers into 16 with 4 numbers each as follows.

Now, just place each group of 4 numbers according to the numerical order of 4x4 magic square and the patterns of Mei's Square of 8x8. An 8x8 magic square will then be made as follow:

16

3	2	59	58	39	38	31	30
1	4	57	60	37	40	29	32
18	19	54	55	42	43	14	15
20	17	56	53	44	41	16	13
62	63	10	11	22	23	34	35
64	61	12	9	24	21	36	33
47	46	7	6	27	26	51	50
45	48	5	8	25	28	49	52

G) 9x9 Magic Square

Method 1 – DMM: There are 12 cells you can start with 16 directions shown on left. A 9x9 magic square has been constructed by DMM on right.

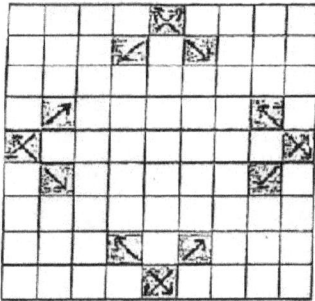

50	81	72	74	4	15	26	28	39
60	71	73	3	14	25	36	38	49
70	81	2	13	24	35	37	48	59
80	1	12	23	34	45	47	58	69
9	11	22	33	44	46	57	68	79
10	21	32	43	54	56	67	78	8
20	31	42	53	55	66	77	7	18
30	41	52	63	65	76	6	17	19
40	51	62	64	75	5	16	27	29

Method 2 – KMM: KMM works fine with 5x5 and 7x7 but not with 9x9. Not even one 9x9 magic square can be made by this method. That is echoed by Mr. W.S Andrew, who foresees the problem as described in his book **Magic Square and Cubes**. WHY? Nobody can give the answer until nearly a century later. It is because 9 is not a prime number. KMM only works with odd prime numbers. Since 9 is not a prime which means it has factors other than 1 and itself, a method called COMPOSITE has been developed to make 9x9 magic squares as follows:

Method 3 – COMPOSITE: Please note that the Composite Method is not limited to non-prime odd numbers but even numbers which have factors other than 2. In order to make 9x9 magic squares, the following steps are employed:

a.) Group the numbers from 1 to 81 into nine sets with 9 numbers each.

(1) 1,2,3,4,5,6,7,8,9　　(2) 10,11,12,13,14,15,16,17,18　(3) 19,20,21,22,23,24,25,26,27
(4) 28,29,30,31,32,33,34,35,36　(5) 37,38,39,40,41,42,43,44,45　(6) 46,47,48,49,50,51,52,53,54
(7) 55,56,57,58,59,60,61,62,63　(8) 64,65,66,67,68,69,70,71,72　(9) 73,74,75,76,77,78,79,80,81.

b.) Form 9 3x3 magic squares with 9 sets of numbers above:

	1				2				3		
	6	7	2		15	16	11		24	25	20
	1	5	9		10	14	18		19	23	27
	8	3	4		17	12	13		26	21	22

	4				5				6		
	33	34	29		42	43	38		51	52	47
	28	32	36		37	41	45		46	50	54
	35	30	31		44	39	40		53	48	49

	7				8				9		
	60	61	56		69	70	65		78	79	74
	55	59	63		64	68	72		73	77	81
	62	57	58		71	66	67		80	75	76

c.) Place all 9 3x3 magic squared positioned as shown below. Then, a 9x9 magic square is created.

51	52	47	60	61	56	15	16	11
46	50	54	55	59	63	10	14	18
53	48	49	62	57	58	17	12	13
6	7	2	42	43	38	78	79	74
1	5	9	37	41	45	73	77	81
8	3	4	44	39	40	80	75	76
69	70	65	24	25	20	33	34	29
64	68	72	19	23	27	28	32	36
71	66	67	25	21	22	35	30	31

H) 10x10 Magic Square

Method 1 – MSM: Before applying MSM, let's group from 1 to 100 into 25 sets with 4 numbers each

(1) 1,2,3,4   (2) 5,6,7,8   (3) 9,10,11,12 (4) 13,14,15,16 (5) 17,18,19,20
(6) 21,22,23,24 (7) 25,26,27,28 (8) 29,30,31,32 (9) 33,34,35,36 (10) 37,38,39,40
(11) 41,42,43,44 (12) 45,46,47,48 (13) 49,50,51,52 (14) 53,54,55,56 (15) 57,58,59,60
(16) 61,62,63,64 (17) 65,66,67,68 (18) 69,70,71,72 (19) 73,74,75,76 (20) 77,78,79,80
(21) 81,82,83,84 (22) 85,86,87,88 (23) 89,90,91,92 (24) 93,94,95,96 (25) 97,98,99,100

Now, we need a 5x5 magic square as the BASE and a Mei's Square for 10x10. See below:

17	24	1	8	15
23	5	7	14	16
4	6	13	20	22
10	12	19	21	3
11	18	25	2	9

Place 25 sets of 4 numbers in numerical order as shown in 5x5 magic square corresponding the 25 patterns of Mei's Square for 10x10 into a blank 10x10 square which will become magic.

66	65	94	96	3	4	30	32	58	57
67	68	95	93	2	1	31	29	59	60
90	89	18	20	27	28	54	53	62	64
91	92	19	17	26	25	55	56	63	61
14	13	22	24	51	52	78	80	86	85
15	16	23	21	50	49	79	77	87	88
38	37	46	48	75	76	82	81	10	12
39	40	47	45	74	73	83	84	11	9
42	41	70	72	99	100	6	8	34	33
43	44	71	69	98	97	7	5	35	36

I) 11x11 Magic Square

Method One – DMM: There are only 40 cells you can start with 44 directions as shown below:

76	78	91	104	117	9	22	24	37	50	63
88	90	103	116	8	21	23	36	49	62	75
89	102	115	7	20	33	35	48	61	74	87
101	114	6	19	32	34	47	60	73	86	99
113	5	18	31	44	46	59	72	85	98	100
4	17	30	43	45	58	71	84	97	110	112
16	29	42	55	57	70	83	96	109	111	3
28	41	54	56	69	82	95	108	121	2	15
40	53	66	68	81	94	107	120	1	14	27
52	65	67	80	93	106	119	11	13	26	39
64	77	79	92	105	118	10	12	25	38	51

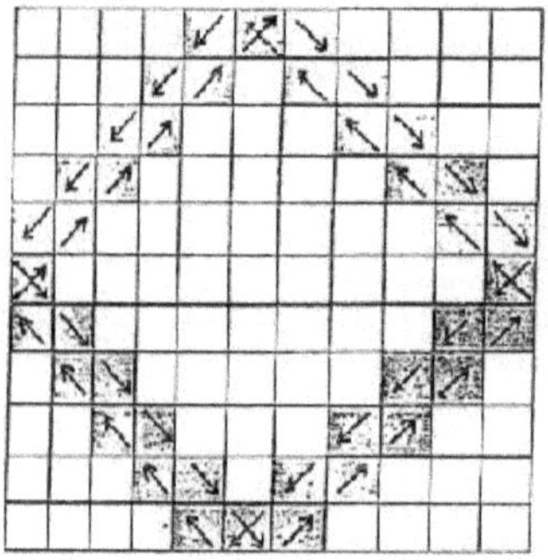

Method TWO – KMM: Because 11 is a prime, you have the luxury to start any cell with any direction by applying this method. It results at least 14,641 11x11 magic squares are formed.

1	26	51	76	90	115	19	44	58	83	108
13	38	63	88	102	6	31	45	70	95	120
25	50	75	89	114	18	43	57	82	107	11
37	62	87	101	5	30	55	69	94	119	12
49	74	99	113	17	42	56	81	106	10	24
61	86	100	4	29	54	68	93	118	22	36
73	98	112	16	41	66	80	105	9	23	48
85	110	3	28	53	67	92	117	21	35	60
97	111	15	40	65	79	104	8	33	47	72
109	2	27	52	77	91	116	20	34	59	84
121	14	39	64	78	103	7	32	46	71	96

J) 12x12 Magic Square

Method One – MSM: Because 12 = 2x5, you need a Mei's Square for 12x12 and a 6x6 magic square as the BASE as shown below.

11	8	25	31	15	21
2	5	34	28	24	18
27	33	23	20	1	7
36	30	14	17	10	4
22	19	3	9	26	32
13	16	12	6	35	29

Now, you have to arrange 1-144 numbers into groups of 4 numbers each as follows:

1) 1,2,3,4	2) 5,6,7,8	3) 9,10,11,12,	4) 13,14,15,16
5) 17,18,19,20	6) 21,22,23,24	7) 25,26,27,28	8) 29,30,31,32
9) 33,34,35,36	10) 37,38,39,40	11) 41,42,43,44	12) 45,46,47,48
13) 49,50,51,52	14) 53,54,55,56	15) 57,58,59,60	16) 61,62,63,64
17) 65,66,67,68	18) 69,70,71,72	19) 73,74,75,76	20) 77,78,79,80
21) 81,82,83,84	22) 85,86,87,88	23) 89,90,91,92	24) 93,94,95,96
25) 97,98,99,100	26) 101,102,103,104	27) 105,106,107,108	28) 109,110,111,112
29)113,114,115,116	30) 117,118,119,120	31) 121,122,123,124	32) 125,126,127,128
33)129,130,131,13	34) 133,134,135,136	35) 137,138,139,140	36) 141,142,143,144

Position the 36 groups of 4 numbers into the 12x12 empty square corresponding the order in the BASE, a 6x6 magic squareandthepatternsofMei'sSquarefor12x12magic square is created as shown below:

44	41	32	29	100	97	124	121	60	57	84	81
42	43	30	31	98	99	122	123	58	59	82	83
8	5	17	20	136	133	112	109	93	96	72	69
6	7	18	19	134	135	110	111	94	95	70	71
105	108	132	129	89	92	77	80	4	1	25	28
106	107	130	131	90	91	78	79	2	3	26	27
141	144	120	117	53	56	65	68	40	37	13	16
142	143	118	119	54	55	66	67	38	39	14	15
88	85	73	76	12	9	36	33	101	104	128	125
86	87	74	75	10	11	34	35	102	103	126	127
52	49	64	61	48	45	24	21	140	137	116	113
50	51	62	63	46	47	22	23	138	139	114	115

Method Two – Composite Method: There are two ways to make 12x12 magic square by this method as follows:

6	7	2
1	5	9
8	3	4

Method Two – A) Using a 3x3 magic square as the BASE Grouping 1 to 144 into 9 sets with 16 numbers each as follows.

1.) 1,2,3,4,5,6,7,8,9,10,11,12,13,14,15,16

2.) 17,18,19,20,21,22,23,24,25,26,27,28,29,30,31,32

3.) 33,34,35,36,37,38,39,40,41,42,43,44,45,46,47,48

4.) 49,50,51,52,53,54,55,56,57,58,59,60,61,62,63,64

5.) 65,66,67,68,69,70,71,72,73,74,75,76,77,78,79,80

6.) 81,82,83,84,85,86,87,88,89,90,91,92,93,94,95,96

7.) 97,98,99,100,101,102,103,104,105,106,107,108,109,110,111

8.) 113,114,115,116,117,118,119,120,121,122,123,124,125,1
26,127,128

9.) 129,130,131,132,133,134,135,136,137,138,139,140,141,1
42,143,144

Each of the 9 sets forms a 4x4 magic square as follows:

Once 9 4x4 magic squares are formed, place them into 9 cells of the BASE, a selected 3x3 magic square matching their corresponding numbers. A 12x12 square is then created.

81	95	94	84	97	111	110	100	17	31	30	20
92	86	87	89	108	102	103	105	28	22	23	25
88	90	91	85	104	106	107	101	24	26	27	21
93	83	82	96	109	99	98	112	29	19	18	32
1	15	14	4	85	79	78	68	129	143	142	132
12	6	7	9	76	70	71	73	140	134	135	137
8	10	11	5	72	74	75	69	136	138	139	133
13	3	2	16	77	67	66	80	141	131	130	144
113	127	126	116	33	47	46	36	49	63	62	52
124	118	119	121	44	38	39	41	60	54	55	57
120	122	123	117	40	42	43	37	56	58	59	53
125	115	114	128	45	35	34	48	61	51	50	64

Method Two – B) Using a 4x4 magic square as the BASE and grouping numbers 1 – 144 into sets with 9 numbers each as follows:

1) 1,2,3,4,5,6,7,8,9

2) 10,11,12,13,14,15,16,17,18

3) 19,20,21,22,23,24,25,26,27

4) 28,29,30,31,32,33,34,35,36

5) 37,38,39,40,41,42,43,44,45

6) 46,47,48,49,50,51,52,53,54

8) 64,65,66,67,68,69,70,71,72

10) 82,83,84,85,86,87,88,89,90

12) 100,101,102,103,104,105,106,107,108

14) 118,119,120,121,122,123,124,125,126

16) 136,137,138,139,140,141,142,143,144

1	15	14	4
12	6	7	9
8	10	11	5
13	3	2	16

7) 55,56,57,58,59,60,61,62,63

9) 73,74,75,76,77,78,79,80,81

11) 91,92,93,94,95,96,97,98,99

13) 109,110,111,112,113,114,115,116,117

15) 127,128,129,130,131,132,133,134,135

Each of the above 16 sets forms a 3x3 magic square as follows:

1

6	7	2
1	5	9
8	3	4

2

15	16	11
10	14	18
17	12	13

3

24	25	20
19	23	27
26	21	22

4

33	34	29
28	32	36
35	30	31

5

42	43	38
37	41	45
44	39	40

6

51	52	47
46	50	54
53	48	49

7

60	61	56
55	59	63
52	57	58

8

69	70	65
64	68	72
71	66	67

9

78	79	74
73	77	81
80	75	76

10

87	88	83
82	86	90
89	84	85

11

96	97	92
91	95	99
98	93	94

12

105	106	101
100	104	108
107	102	103

13

114	115	110
109	113	117
116	111	112

14

123	124	119
118	122	126
125	120	121

15

132	133	128
127	131	135
134	129	130

16

141	142	137
136	140	144
143	138	139

Once 16 3x3 magic squares are formed, place them into 16 cells of the 4x4 magic square (BASE) matching their corresponding numbers. A 12x12 magic square is then created.

6	7	2	132	133	128	123	124	119	33	34	29
1	5	9	127	131	135	118	122	126	28	32	36
8	3	4	134	129	130	125	120	121	35	30	31
105	106	101	51	52	47	60	61	56	78	79	74
100	104	108	46	50	54	55	59	63	73	77	81
107	102	103	53	48	49	62	57	58	80	75	76
69	70	65	87	88	83	96	97	92	42	43	38
64	68	72	82	86	90	91	95	99	37	41	45
71	66	67	89	84	85	98	93	94	44	39	40
114	115	110	24	25	20	15	16	11	141	142	137
109	113	117	19	23	27	10	14	18	136	140	144
116	111	112	26	21	22	17	12	13	143	138	139

Beyond 12x12 (Continued)

456	457	452	465	466	461	420	421	416	537	538	533	546	547	542	501	502	497	132	133	128	141	412	137	96	97	
451	455	459	460	464	468	415	419	423	532	536	540	541	545	549	496	500	504	127	131	135	136	140	144	91	95	
458	453	454	467	462	463	422	417	418	539	534	535	548	543	544	503	498	499	134	129	130	143	138	139	98	93	
411	412	407	447	448	443	483	484	479	492	493	488	528	529	524	564	565	560	87	88	83	123	124	119	159	16	
406	410	414	442	446	450	478	482	486	487	491	495	523	527	531	559	563	567	82	86	90	118	122	126	154	15	
413	408	409	449	444	445	485	480	481	494	489	490	430	525	526	566	561	562	89	84	85	125	120	121	161	15	
474	475	470	429	430	425	438	439	434	555	556	551	510	511	506	519	520	515	150	151	146	105	106	101	114	115	
469	473	477	424	428	432	433	437	441	550	554	558	505	509	513	514	518	522	145	149	153	100	104	108	109	113	
476	471	472	431	426	427	440	435	436	557	552	553	512	507	508	521	516	517	152	147	148	107	102	103	116	111	
51	52	47	60	61	56	15	16	11	375	376	371	384	385	380	339	340	335	699	700	695	708	709	704	663	66	
46	50	54	55	59	63	10	14	18	370	374	378	379	383	387	334	338	342	694	698	702	703	707	711	658	66	
53	48	49	62	57	38	78	79	74	377	372	373	386	381	382	341	336	337	701	696	697	710	705	706	665	66	
6	7	2	41	43	38	78	79	74	330	331	326	366	367	362	402	403	398	654	655	650	690	691	686	726	72	
1	5	9	37	41	45	73	77	81	325	329	333	361	365	369	397	401	405	649	653	657	685	689	693	721	72	
8	3	4	44	39	40	80	75	76	332	327	328	368	363	364	404	399	400	656	651	652	692	687	688	728	72	
69	70	65	24	25	20	33	34	29	393	394	389	348	349	344	357	358	353	717	718	713	672	673	668	681	68	
64	68	72	19	23	27	28	32	36	388	392	396	343	347	351	352	356	360	712	716	720	667	671	675	576	68	
71	66	67	67	26	22	35	30	31	395	390	391	950	345	346	359	354	355	719	714	715	674	669	670	683	67	
618	619	614	627	628	623	582	583	578	213	214	209	222	223	218	177	178	173	294	295	290	303	304	299	258	25	
613	617	621	622	626	630	577	581	585	208	212	216	217	221	225	172	176	180	289	293	297	298	302	306	253	25	
620	615	616	629	624	625	584	579	580	215	210	211	224	219	220	179	174	175	296	291	292	305	300	301	250	25	
573	574	569	609	610	605	645	646	641	168	169	164	204	205	200	240	241	236	249	250	245	285	286	281	321	32	
568	572	576	604	608	612	640	644	648	163	167	171	199	203	207	235	239	243	244	248	252	280	284	288	316	32	
575	570	571	611	606	607	647	642	643	170	165	166	206	201	202	242	237	238	251	246	247	287	282	283	323	31	
636	637	632	591	592	587	600	601	596	231	232	227	186	187	182	195	196	191	312	313	308	267	268	263	276	27	
631	635	639	586	590	594	595	599	603	226	230	234	181	185	189	190	194	198	307	311	315	262	266	270	271	27	
638	633	634	593	588	589	602	597	598	233	228	229	188	183	184	197	132	193	314	309	310	269	264	265	278	27	

27 x 27 Magic Square made by Super Composite Method

K) Magic Squares beyond 12x12

13x13 magic square can be made by KMM 14x14 magic square can be made by MSM

15x15 magic square can be made by Composite Method

24x24 magic square can eb made by MSM or Composite Method with 4 different BASES (3, 4, 6, 8)

|
|
|
|

27x27 magic square can be made by Super Composite Method as follows. Do not get scared by this gigantic size. You can make it without any difficulties in minutes. All you have to do is to start with a 3x3 magic square. Observe these squares with sizes 3x3, 9x9, 27x27 and see 27 = 3 x 3x 3 that applies the Composite Method not once but twice. For now, do you get the idea?

How long it takes you to write the numbers from 1 to 729 is how long it takes you to make a 27x27 MAGIC SQUARE. IS THAT EASY??

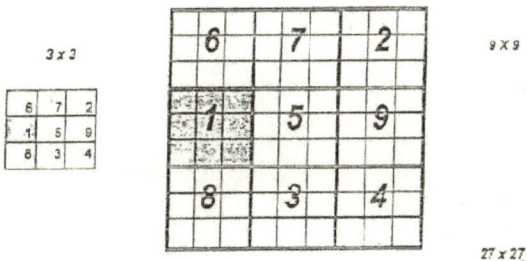

3 x 3

6	7	2
1	5	9
8	3	4

9 X 9

6		7		2	
1		5		9	
8		3		4	

27 x 27

27

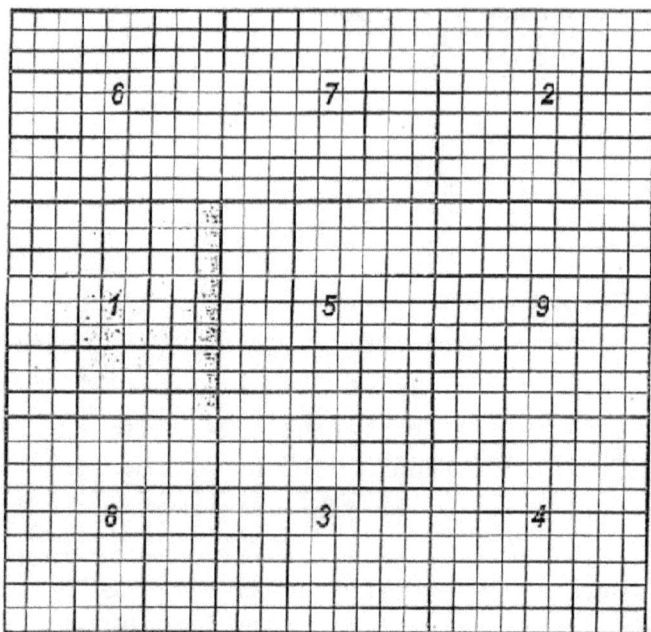

Beyond 12x12 (Continued)

456	457	452	465	466	461	420	421	416	537	538	533	546	547	542	501	502	497	132	133	128	141	142	137	96	97
451	455	459	460	464	468	415	419	423	532	536	540	541	545	549	496	500	504	127	131	135	136	140	144	91	95
458	453	454	467	462	463	422	417	418	539	534	535	548	543	544	503	498	499	134	129	130	143	138	139	98	93
411	412	407	447	448	443	483	484	479	492	493	488	528	529	524	564	565	560	87	88	83	123	124	119	159	16
406	410	414	442	446	450	478	482	486	487	491	495	523	527	531	559	563	567	82	86	90	118	122	126	154	15
413	408	409	449	444	445	485	480	481	494	489	490	430	525	526	566	561	562	89	84	85	125	120	121	161	15
474	475	470	429	430	425	438	439	434	555	556	551	510	511	506	519	520	515	150	151	146	105	106	101	114	115
469	473	477	424	428	434	433	437	441	550	554	558	505	509	513	514	518	522	145	149	153	100	104	108	109	113
476	471	472	431	426	427	440	435	436	557	552	553	512	507	508	521	516	517	152	147	148	107	102	103	116	111
51	52	47	60	61	56	15	16	11	375	376	371	384	385	380	339	340	335	699	700	695	708	709	704	663	66
46	50	54	55	59	63	10	14	18	370	374	378	379	383	387	334	338	342	694	698	702	703	707	711	658	66
53	48	49	62	57	38	78	79	74	377	372	373	386	381	382	341	336	337	701	696	697	710	705	706	665	66
6	7	2	42	43	38	78	79	74	330	331	326	366	367	362	402	403	398	654	655	650	690	691	686	726	72
1	5	9	37	41	45	73	77	81	325	329	333	361	365	369	397	401	405	649	653	657	685	689	693	721	72
8	3	4	44	39	40	80	75	76	332	327	328	368	363	404	399	399	400	656	651	652	692	687	688	728	72
69	70	65	24	25	20	33	34	29	393	394	389	348	349	344	357	358	353	717	718	713	672	673	668	681	68
64	68	72	19	23	27	28	32	36	388	392	396	343	347	351	352	356	360	715	716	720	667	671	675	576	68
71	66	67	67	26	22	35	30	31	395	390	391	350	345	346	359	354	355	719	714	715	674	669	670	683	67
618	619	614	627	628	623	582	583	578	213	214	209	222	223	218	177	178	173	294	295	290	303	304	299	258	25

613	617	621	622	626	630	577	581	585	208	212	216	217	221	225	172	176	180	289	293	297	298	302	306	253	25
620	615	616	629	624	625	584	579	580	215	210	211	224	219	220	179	174	175	296	291	292	305	300	301	250	25
573	574	569	609	610	105	645	646	641	168	169	164	204	205	200	240	241	236	249	250	245	285	286	281	321	32
568	572	576	604	608	612	640	644	648	163	167	171	199	203	207	235	239	243	244	248	252	280	284	288	316	32
575	570	571	611	606	607	647	642	643	170	165	166	206	201	202	242	237	238	251	246	247	287	282	283	323	31
636	637	632	591	592	587	600	601	596	231	232	227	186	287	182	195	196	191	312	313	308	267	268	263	276	27
631	635	639	586	590	594	595	599	603	226	230	234	181	185	189	190	194	198	307	311	315	262	266	270	270	27
638	633	634	593	588	589	602	597	598	233	228	229	188	183	184	197	192	193	314	309	310	269	264	265	278	27

27 x 27 Magic Square made by Super Composite Method

L) Summary

This chapter is hereby concluded with the table shown below:

METHODS	SIZES OF MAGIC SQUARE																
	4x4	5x5	6x6	7x7	8x8	9x9	10x10	11x11	12x12	13x13	14x14	16x16	17x17	18x18	19x19	20x20	21x21
D or K	·		·		·		·		·		·		·		·		·
R / C	·	·															
Composite				·				·				·		·		·	·
MSM			·		·		·		·		·	·		·		·	

By observing the table, one should find out the importance of the invention of MSM in creating even number magic squares. It not only fills up the gap in the making any magic square, but also enables us to construct any size of magic square with only paper and pen. The next chapter will illustrate in details this fascinating discovery.

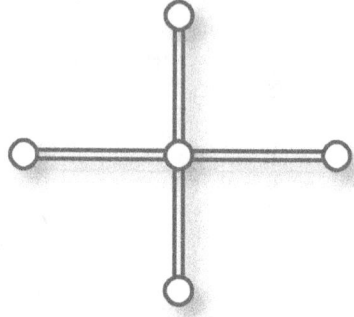

III

The Discovery of Mei's Square Method

The MSM has three major elements: The BASE, the Arrangement of Numbers and Mei's Square. The examples of MSM in making even number magic squares are describe in the previous chapter.

A) The BASE is a magic square used to accommodate a series of sets of 4 numbers according to numerical order within the magic square, The BASE. A general formula is developed for the size of a magic square to be used as The BASE in MSM as follows:

In making an AxA even number magic square, The BASE needs to be a A/2 x A/2 magic square

For 6x6, a 3x3 magic square is used as BASE.
For 8x8, a 4x4 magic square is used as BASE.
For 10x10, a 5x5 magic square is used as BASE.
For 12x12, a 6x6 magic square is used as BASE.

B) The Arrangement of Numbers has three different formats as below:

INTERVAL – It is defined as the difference between two consecutive numbers. In MSM, the intervals of 4 numbers in each group must stay constant, The following examples will show you how the 36 numbers are arranged into 9 groups with 3 different intervals.

INTERVAL - 1	INTERVAL - 3	INTERVAL - 9
1) 1,2,3,4	1) 1,4,7,10	1) 1,10,19,28
2) 5,6,7,8	2) 2,5,8,11	2) 2,11,20,29
3) 9,10,11,12	3) 3,6,9,12	3) 3,12,21,30
4) 13,14,15,16	4) 13,16,19,22	4) 4,13,22,31
5) 17,18,19,20	5) 14,17,20,23	5) 5,14,23,32
6) 21,22,23,24	6) 15,18,21,24	6) 6,15,24,33
7) 25,26,27,28	7) 25,28,31,34	7) 7,16,25,34
8) 29,30,31,32	8) 26,29,32,35	8) 8,17,26,35
9) 33,34,35,36	9) 27,30,33,36	9) 9,18,27,36

ORDER – The order of 9 groups of 4 numbers can be reversed or alternated

1) 33,34,35,36	1) 1,2,3,4
2) 29,30,31,32	2) 13,14,15,16
3) 25,26,27,28	3) 25,26,27,28
4) 21,22,23,24	4) 5,6,7,8
5) 17,18,19,20	5) 17,18,19,20
6) 13,14,15,16	6) 29,30,31,32
7) 9,10,11,12	7) 9,10,11,12
8) 5,6,7,8	8) 21,22,23,24
9) 1,2,3,4	9) 33,34,35,36.

SEQUENCE – The sequence of the 4 numbers of each 9 groups can be ascending or descending.

1,2,3,4 or 4,3,2,1; 2,5,8,11 or 11,8,5,2

C) Mei's Square is an important element of MSM in making even number magic squares. To construct a 6x6 magic square, a 3x3 magic square as The BASE and a Mei's Square for 6x6 are needed. A Mei's Square for 6x6 has 9 patters which correspond the total numbers in the 3x3 magic square served as The BASE. The patterns can be the same in some, or completely different, as long as they lead to a successful construction of a magic square. There are 24 movements of pattern for a set of 4 numbers as shown below: (The open end is where to start and the bar is where to stop)

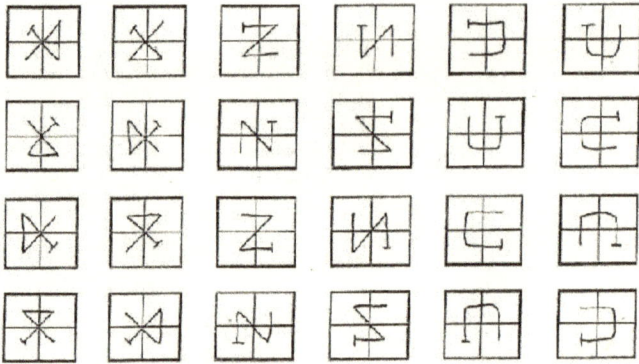

The following examples show how these patterns are applied to form Mei's Squares:

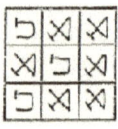

Mei's Square for 6x6
Example # 1

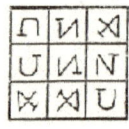

Mei's Square for 6x6
Example # 2

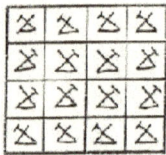

Mei's Square for 8x8
Example # 1

Mei's Square for 8x8
Example # 2

Please note that both Example 1s have only two patterns while the other two have all different patterns.

The discovery of Mei's Square Method gives an easy way to make even number magic squares. The key of this method is the development of Mei's Square which shows the movement of 4 numbers in patterns. The following steps show you how the Mei's Square for 6x6 is discovered:

STEP 1: Let's have a 3x3 magic square randomly.

STEP 2: Using interval of 1, group numbers 1 – 36 into 9 each with 4 numbers.

1) 1,2,3,4	2) 5,6,7,8	3) 9,10,11,12
4) 13,14,15,16	5) 17,18,19,20	6) 21,22,23,24
7) 25,26,27,28	8) 29,30,31,32	9) 33,34,35,36.

STEP 3: Apply pattern Z in placing above 9 groups of 4 numbers into a blank 6x6 square in corresponding positions as in 3x3 magic square.

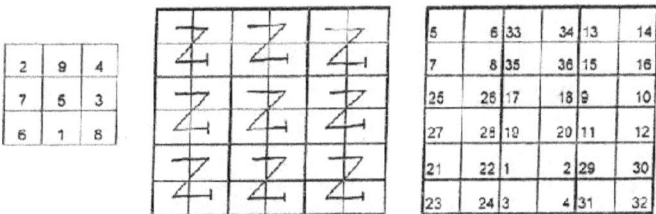

STEP 4: Add up each row, column and diagonal with results shown below:

	+3	-3	+3	-3	+3	-3		
111	108	114	108	114	108	114	111	
5	6	33	34	13	14		105	+6
7	8	35	36	15	16		117	-6
25	26	17	18	9	10		105	+6
27	28	19	20	11	12		117	-6
21	22	1	2	29	30		105	+6
23	24	3	4	31	32		117	-6

STEP 5: Of course they are not added up to 111 except two diagonals.

However, we can fix them. One Step At A Time. Let's take a look at the sum of rows which are either +6 or -6 off the target 111. We can make them all equal to 111 by switching numbers up and down or vice versa in each adjacent rows and make sure the 2 diagonals maintain 11 as shown below.

7 +2	8 -2		36 +2		
5	6	33	34	13	14
5 -2	8 -2	34 -2			
7	8	35	36	15	16
		19 +2	20 +2	11 +2	
25	26	17	18	9	10
		17	18 -2	9 -2	
27	28	19	20	11	12
				31 +2	32 +2
21	22	1	2	29	30
				29 -2	30 -2
23	24	3	4	31	32

+3 108	−3 114	+3 108	−3 114	+3 108	−3 114
7	8	33	36	13	14
5	6	35	34	15	16
25	26	19	20	11	10
27	28	17	18	9	12
21	22	3	2	31	32
23	24	1	4	29	30

STEP 6: Now, let's work on columns by switching numbers left to right or vice versa in each adjacent ones. Then a 6x6 magic square has been created as follows.

8 +1 7	7 −1 8	36 +3 33	33 −3 36	13	14
6 +1 5	5 −1 6	35	34	15	16
26 +1 25	25 −1 26	19	20	11	10
27	28	17	18	12 +3 9	9 −3 12
21	22	3	2	31	32
23	24	1	4	29	30

8	7	36	33	13	14
6	5	35	34	15	16
26	25	19	20	11	10
27	28	17	18	12	9
21	22	3	2	31	32
23	24	1	4	29	30

6x6 Magic Square

STEP 7: This time we reverse the steps. We are getting a Mei's Square for 6x6 by copying the patterns. Then, you can try the other possibilities in Steps 5 & 6 to haave more Mei's Squares.

3 adaditionaal Mei's Squre for 6x6 kcana be obtained by rotating the above:

The MSM can cover all in making even number magic squares except 4x4. There is one printed in <u>Games, Ancient and Oriented</u> by Edward Falkener, published by Longmans Green & Co., London and New York, 1892. How he came up with that magic square below is unknown, but it is easily covered by one of 40,976 Mei's Square for 6x6 found so far.

32	31	1	3	21	23
29	30	4	2	24	22
9	11	20	19	25	27
12	10	17	18	28	26
16	15	33	35	5	7
13	14	36	34	8	6

The number of 6x6 magic squares, constructed by MSM can be shown in the following formula:

# of 6x6 M.S. = # of 3x3 M.S x # of intervals x # of Sequences x Rule of Family of 8 x # of Mei's Square for 6x6

$$= 8 \quad x \quad 3 \quad x \quad 3 \quad x \quad 2 \quad x \quad 8 \quad x \quad 40.,976$$

$$= 47,204,352$$

The 47,204,352 is a conservative figure because the research of Mei's Square for 6x6 is wide open. The next even number of magic square 8x8 can be constructed by MSM in an astonishing quantity. This mass production paves the way of many valuable applications of magic square.

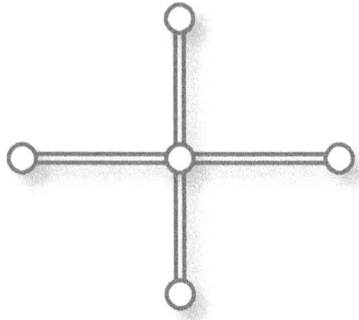

IV

The Applications of Magic Squares

Dr. Paul Carcus, a well-known mathematician in late 19th century, has said, "Magic Squares have no immediate practical use. "He is indeed right because nothing you can do with so limited quantity of magic square of all sizes in his time. It changes until a century later, methods of mass production for magic square of all sizes have been discovered, attributing the birth if one of the outstanding application of magic square: MAGIC SQUARE PUZZLE. However, this chapter will start with the following application.

A) The Extension of Knight's Move Method:

An interesting property of Knight's Move Method is being discovered. Let's begin with a 5x5 magic square created by the KMM and put the 5x5 together with its copy as shown in Diagram A. Each of all diagonals adds up to 65 which happenes to be the same amount of each row and column of the 5x5 magic square.

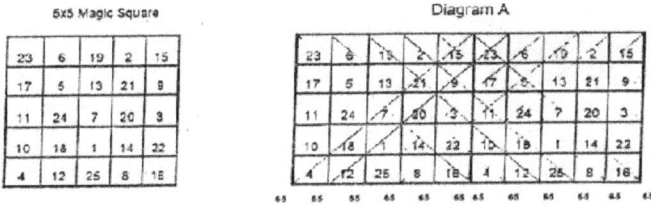

5x5 Magic Square

23	6	19	2	15
17	5	13	21	9
11	24	7	20	3
10	18	1	14	22
4	12	25	8	16

Diagram A

This property enables us to have the following two applications:

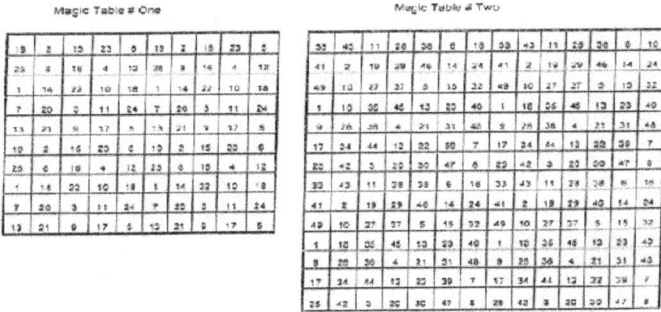

Magic Table # One

Magic Table # Two

In Magic table # One and Two, any 5 and 7 numbers respectively succeeding on straight line vertically, horizontally or diagonally, add up to a same amount equals to 65 and 175 respectively. In a closer look, there are 4 same 5x5 and 7x7 magic squares positioned in Magic Tables One and Two respectively.

The other application is an amusing one. We can have an odd prime number magic square custom-made which means a desired number can be at a desired location. The following example shows the steps to set the desired number 19 at the center of the square:

STEP 1: Construct a 5x5 magic square randomly with Knight's Move Method

12	25	8	16	4
18	1	14	22	10
24	7	20	3	11
5	13	21	9	17
6	19	2	15	23

STEP 2: Rotate the column that contains the number 19 to the middle.

4	12	25	8	16
10	18	1	14	22
11	24	7	20	3
17	5	13	21	9
23	6	19	2	15

STEP 3: Rotate the row that contains the number 19 to the middle. Then you get a custom-made 5x5 magic square.

11	24	7	20	3
17	5	13	21	9
23	6	19	2	15
4	12	25	8	16
10	18	1	14	22

B) MAGIC SQUARE PUZZLE: It is defined as a magic square with missing numbers. To solve the MSP is to find the missing numbers in the right positions and make the square magic again. The following are the elements, concepts and methods in solving a magic square puzzle:

ELEMENTS

a) BASE: is the magic square used for the puzzle.

b) FORMAT: is the arrangement of the missing numbers. It can be random, symmetrical or in block or blocks.

c) The number of missing numbers is standardized by the use of the formula: For a AxA magic square puzzle, the total of missing numbers equals to 2A.

SIZE	# MISSING
3x3	6
4x4	8
5x5	10
6x6	12
7x7	14
8x8	16

CONCEPTS

a) Different sizes of magic square have a fixed amount of numbers. It all starts with the number 1, ends with the number 9 for 3x3, 16 for 4x4, 25 for 5x5, 36 for 6x6, 49 for 7x7, 64 for 8x8.

b) No number is ever used more than once.

c) 7 additional magic squared can be obtained by switching rows, columns or both. This is called THE RULE OF FAMILY OF 8. The logical way to prove it is as follows:

There are 4 corner numbers in each square which can be counted in a row column. It turns to be 8 different positioned magic squares.

	1				2				3				4	
2	9	4		4	9	2		6	1	8		8	1	6
7	5	3		3	5	7		7	5	3		3	5	7
6	1	8		8	1	6		2	9	4		4	9	2

	5				6				7				8	
2	7	6		4	3	8		6	7	2		8	3	4
9	5	1		9	5	1		1	5	9		1	5	9
4	3	8		2	7	6		8	3	4		6	7	2

d) There are two basic theorems in magic square with proof as follows:

1. Magic Square theorem I: The sum of all numbers in a AxA magic square equals to A2(1 + A2) / 2, where A is positive integer. ROOF: The sum of all numbers in a AXA magic square will be

$$S = 1+2+3+ \ldots\ldots\ldots\ldots\ldots\ldots\ldots\ldots\ldots\ldots\ldots\ldots + (A2 - 1) + A2$$

CASE One: When A is a positive even integer, we observe these pairs:

1 & A2, 2 & (A2–1) …………….and they all equal such asa
1 + A2=2+ (A2-1) = so on.

They are exact A2/2 pairs. Therefore, the sum will be
S = A2 (1 + A2) / 2

CASE TWO: When A is a positive odd integer, we observe that there is a middle number which always equals to (1 + A2) / 2

The rest are paired up and their sums are equal. There Will be $(1 + A) / 2 - 1$ pairs. Now the sum will be:

$S = (1 + A2) / 2 + (1 + A2) \times [(1 + A2) / 2 - 1]$
$= (1 + A2) / 2 + (1 + A2) \times (A2 - 1) / 2$
$= (1 + A2) / 2 + (A4 - 1) / 2$
$= (AxA + A4) / 2$
$= A2 (1 + A2) / 2$

We conclude the sum of all numbers in a AxA magic square equals to $A2 (1 + A2) / 2$, where A is a positive integer.

2. Magic Square Theorem II: The sum of numbers on each row, column and diagonal adds up to a same amount equals to $A (1 + A2) / 2$ in a AxA magic square, where A is a positive integer.

PROOF: There are A rows and columns in a AxA magic square. Based on Theorem I, each row and column will equal to

$$A2 (1 + A2) / 2 \div A = A (1 + A2) / 2$$

So does each diagonal by the definition of a magic square.

METHODS

There are many ways to solve a MSP. Once can solve a simple puzzle by jiggling those missing numbers over empty cells. The following methods are used to solve all kind of magic square puzzles:

a) Subtraction   b) Pair-Up

c) Simplifying by Elimination

d) Simplifying by Contradiction

e) Assumption   f) On-Your-Own

EXAMPLE ONE: Solve the 4x4 MSP.

12	B	16	G
A	11	6	H
2	14	C	D
13	4	E	F

1) Find out the missing numbers as follows: 1, 3, 5, 7, 8, 9, 10, 15

2) Use A, B, C, D, E, F, G, & H for missing numbers and place them in empty cells.

3) The same sum for 4x4 is 34. Then, A = 34 – 12 – 2 – 13 =7, B= 34 – 11 – 4 = 5, G= 34 – 12 – B – 16 = 1, H = 34 – A – 11 – 6 =10 (by subtraction)

4) Since the rest cannot be found by subtraction, Pair-Up method will be used. C +D = 34 – 2 – 14 = 18, Choose a pair of rest of missing numbers that satisfies the equation above. Only the pair of 3 and 15 firs. Repeat the same procedures to get C + E = 34 -16 – 6 = 12 = 9 + 3. Compare the two equations,

C + D = 3 + 15 and C + E = 3 + 9 C = 3, D = 15, E = 9.

5) F = 34 – D – 1 – 10 = 8 (by substraction)

EXAMPLE TWO: Solve 6x6 MSP

17	30	22	8	21	13
10	23	G	H	32	9
6	A	I	J	F	11
26	B	K	L	E	31
28	5	C	D	14	27
24	7	29	15	16	20

1) Identify the missing numbers: 1, 2, 3, 4, 18, 19, 25, 33, 34, 35 & 36.

2) Place A – L in the empty cells.

3) No answer can be obtained just by subtraction but we do have sums of the following pairs: (Apply Pair-Up Method)

A+B=46=12+34 ,   E+F=28–3=25

G+F=37=1+36=2+35=(3+34)=4+33=18+19

C+D=37=1+36=2+35=(3+34)=4+33=18+19
I +L=37=1+36=2+35=(3+34)=4+33=18+19
J+K=37=1+36=2+35=(3+34)=4+33=18+19

(3+34) is eliminated for 3 and 34 are possible answers of A,B,E or F.

4) Pick the two simplest pairs to set up 4 set of possible answers:

(Apply Assumption Method)

I	II	III	IV
A = 12	A = 34	A = 12	A = 34
B =34	B = 12	B = 34	B = 12
E =3	E = 3	E = 25	E = 25
F =25	F = 25	F = 3	F = 3

5) Set up I + J = 111-6-A-F-11 to test the validity of the 4 possible sets of answers in 4).

I	II	III	IV
I + J = 57	I + J = 35	I + J = 79	I + J =57
No	= (1 + 34)	No	No
*	= 2 + 33	*	*

6) Eliminating 1 + 34, we have I + J = 2 + 33. Compare it with I + L and I + K above, we have 2 possible sets of answers.

(a) I=33, J=2, K=35, L=4
(b) I=2, J=33, K=4, L=35

7) Set G+C=111-22-l-k-29 to test the possible answers above:

G+C = -8   G+C = 54
No   =18+36=(19+35)=(25+34)

8) Simplify by eliminating 19+35 and 25+34 and get G+C=18+36. Apply the Assumption Method again. We do have 2 sets of answers as follows:

A	B	C	D	E	F	G	H	I	J	K	L
34	12	36	1	3	25	18	19	2	33	4	35
34	12	18	19	3	25	36	1	2	33	4	35

EXAMPLE THREE: Solve the 6x6 MSP

35	12	22	8	3	31
10				5	36
24			33		20
17		4			13
19	32				9
6	25	29	15	34	2

1) Identify the missing numbers: 1, 7, 11 ,14, 16, 18, 21, 23, 26, 27, 28, 30.

2) Place A – L in the empty cells.

3) By observation, none of previous methods work this time. However, we do have the following 4 equations with 3 unknowns:

A+D+G=41, E+D+F=34, H+G+I=77, L+F+I =69

4) We have to find the ice-breaking trick by adding them up.

(A+E+H+L) + 2 (D+G+F+I) = 222

35	12	22	8	3	31
10	A	B	C	5	36
24	D	E	33	F	20
17	G	4	H	I	13
19	32	J	K	L	9
6	25	29	15	34	2

5) SubstitutewithA+E+H+L=74,D+F=34-E, G+I=77-H. Then, we have E+H=37 and A+L=37

6) Take E+H=37 = 7+30 = 14+23 = 16+21 = 11+26 and assume 8 sets of possible answers as follows:

E=7 &H=30E=30 & H=7 E=14 & H=23 E=23 & H=14
E=16 & H=21 E=21 & H=16 E=11 & H=26E=26 & H=11

7) Test above the D+F=24-E, G+I=77–H and B+J =56-E. Only E=11 & H=26 are found valid as the rest are eliminated by logical analysis.

8) Now, we assume 6 sets of possible answer from A+L=7+30=14+23=16+21 (11+26 is eliminated) as follows:

A=7 & L=30, A=30 & L=7, A=14 & L=14
A=23 & L=14, A=16 & L=21, A=21 & L=16

9) Test above with B+C=60 – A, D+G=42 – A, J+K=51 – L and F+I=69-L. Only A=14 & L=23 are found valid as the rest are eliminated by logical analysis.

10) Once we have solve A, E, H & L, the rest should come out as follows:

A	B	C	D	E	F	G	H	I	J	K	L
14	18	28	7	11	16	21	26	30	27	1	23

There are factors determining the difficulties of solving magic square puzzles as follows:

1) The Missing Numbers:

The Amount of Missing Numbers: The more numbers missing is harder to solve. The amount of missing numbers in different sizes of MSP is standardized as follows:

SIZE OF MSP # OF MISSING NUMBERS

3X3	6
4X4	8
5X5	10
6X6	12
7X7	14
8X8	16
AxA	2A

The Format of Missing Numbers: The positional arrangement of missing numbers in random is easier to solve than that in blocks and symmetrical orders. If the format is in one block, it will be harder to solve than those in blocks. In a closer look, if the block is situated along the side or sides, it would be easier to solve than that inside the square. The hardest one is when the block is at the center of the square.

The Intervals of Missing Numbers: If they appear in regular intervals such as 1,2,3,4,5,6,7,……..or 1,4,7,10,13,16……… it will be much harder to solve than those in irregular intervals.

2) The Base:

The BASE is a magic square used for the puzzle. If the BASE is constructed by known methods, the MSP might be easier to solve.

With all these factors floating around, we have a very rough time in determining the level of a MSP. The easiest one should be the MSP with random format and the hardest will be the combinations of above factors, such as one block at the center with regular intervals. So the best way to decide the level of a MSP is to solve one by one and then label them with the following rationales.

ELEMENTARY LEVEL: Solved by subtraction and simple PAIR-UP method.

INTERMEDIATE LEVEL: Solved by complicated PAIR-UP method, simple ASSUMPTION method, moderate math analysis and logical reasoning.

ADVANCED LEVEL: Solved by all advanced methods with complicated math analysis and logical reasoning.

SUPER LEVEL: Solved by equations with three or more unknowns or additional properties of magic squares.

C) The Game of 24

There are 24 ways to place 1, 2, 3 & 4 four numbers to four corners of square as follows:

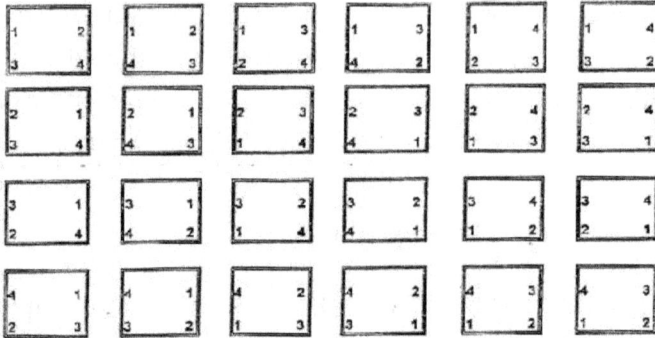

The Game of 24 has two versions:

a) If you can place 9 squares together and add up to 6 numbers vertically, horizontally, or diagonally, equal to 15, you score points. The following are two examples.

4	3	1	3	3	1
2	1	2	4	4	2
1	2	4	3	1	4
3	4	1	2	2	3
3	4	4	1	1	2
2	1	3	2	4	3

2	1	3	2	3	4
4	3	1	4	1	2
2	3	4	3	1	2
4	1	2	1	3	4
1	3	3	4	3	1
2	4	2	1	4	2

b) This one is a little harder. If you can arrange 16 squares and add up 8 numbers vertically, horizontally, or diagonally equal to 20, you score more points. The following are two examples.

2	1	3	4	4	3	1	2
3	4	2	1	1	2	4	3
1	4	2	3	3	2	4	1
3	2	4	1	1	4	2	3
4	1	1	4	3	4	2	1
3	2	2	3	1	2	4	3
1	2	4	3	3	2	2	3
3	4	2	1	4	1	1	4

1	4	4	2	2	4	1	2
2	3	1	3	1	3	4	3
3	1	4	1	4	3	1	3
4	2	3	2	1	2	4	2
2	4	2	1	3	2	4	2
3	1	3	4	4	1	3	1
3	4	1	3	3	1	2	3
2	1	2	4	2	4	1	4

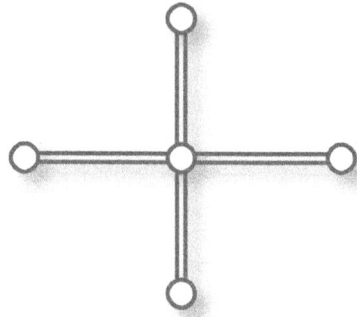

V

Magic Square Now and Beyond

The discovery of R/C and MSM methods fills the gap in constructing any size of magic squares. Especially, the MSM has a mass-productional capacity that contributes millions of combinations and paves the way to the creation of Magic Square Puzzle, one of the outstanding applications of magic squares. The innovative approach of Magic Square Puzzle is to get people of all ages interested in mathematics and improve their mathematical skills through a game setting environment. They will enjoy playing the puzzle and at the same time, they will be learning several mathematical skills, such as problem solving through creative and logical thinking, and simple, moderate & complicated mathematical analysis, as well as algebraic and geometric functions. The object of Magic Square Puzzle is to challenge player to place the missing numbers back to the right cells and make the square magic. Players will experience an emotional and intellectual sense of satisfaction and fulfillment while they fine tune their thinking apparatus.

The magic Square Puzzle has the potential to become a universally popular game while improving the mathematical skills of millions of people. The research of magic square should be continued in order to generate more discoveries and applications in benefit of all mankind.

All quotations are from one and only source below:

MAGIC SQUARE AND CUBE
By: W.S. ANDREW

Solve this gigantic 30x30 magic square puzzle with the same amount of 13,515.

Which means to make each row, column, and diagonal add up to 13,515

Hint: If you know how to make 30x30 magic square, you can solve this in minutes.

Week S001:

Solve the 4x4 MSP by filling in with numbers on the right to make the sum of each row, column, and diagonal equal to 34.

4CC128S001B

				1,2,3,4
7	12	1	14	5,6,7,8
16	3	10	5	

4AA001S001A

	12		14	2,4,6,8,9
	13		11	11,13,15
10		16		
15		9		

Solve the 6x6 MSP by filling in with numbers on the right to make the sum of each row, columns, and diagonal equal to 111.

6AAA114S001B

17	21	31			13	1,2,3,4,5
19	14	27		32		6,7,8,9,10
15	7			25	29	11,12
8	30	4	35	12		
28		36				
24		11	6	16	20	

6AAA105S001A

	28	15	17	19	24	1,2,3,5,9,10
21			30	14	34	18,22,23,26
31	36			27		33,34
35		33	26			
	23	25		32	16	
13		20	22	18	29	

Solve the 8x8 MSP by filling in the numbers on the right to make the sum of each row, column, and diagonal equal to 260.

8A10018001

	24		59	33	56	46		1,2,3,4,5,6,7,8,9
47	58		53		26	36	21	10,11,12,13,14,15,16
51	38	32	41	19		64		
29		18	39	61	44	50		
34	55	45	28		23		60	
	25	35	22	48	57		54	
20		63		52	37	31	42	
62	43	49		30		17	40	

Week S002:

Solve the 4x4 MSP by filling in with numbers on the right to make the sum of each row, column, and diagonal equal to 34.

4BB1263002A

	4	15	
7			12
2			13
	5	10	

1,3,6,8.

9,11,14,16.

4DD024S002B

6			9
	1	14	
	8	11	
3			16

2,4,5,7.

10,12,13,15.

Solve the 6x6 MSP by filling in with numbers on the right to make the sum of each row, columns, and diagonal equal to 111.

6CCC016S002A

		2	6	34	20
	36	1	23	27	
8	21			12	13
33	16			7	29
10	14	9	28		
17	30	22	35		

3,4,5,11,15,

18,19,24,25,

26,31,32.

6BBB072S002B

3	17			21	35
32	1			14	28
7	33			25	6
30	26			12	8
23	10			5	19
16	24			34	15

2,4,9,11,13,

18,20,22,27.

29,31,36.

Solve the 8x8 MSP by filling in the numbers on the right to make the sum of each row, column, and diagonal equal to 260.

8D8002S002

2	59					56	13
16	53	42	19	32	37	58	3
	52	47	6	9	36	63	
	46	49	28	23	62	33	
	11	24	45	34	27	8	
	5	26	35	48	21	10	
41	4	31	54	57	20	15	38
55	30					17	60

1,7,12,14,18,22,25,29,

39,40,43,44,50,51,61,64.

Week S003:

Solve the 4x4 MSP by filling in with numbers on the right to make the sum of each row, column, and diagonal equal to 34.

4AA009S003A

2			13
	5	10	
	4	15	
7			12

1,3,6,8,

9,11,14,16.

4AA023S003B

		6	3
		9	16
11	14		
8	1		

2,4,5,7,

10,12,13,15.

Solve the 6x6 MSP by filling in with numbers on the right to make the sum of each row, columns, and diagonal equal to 111.

6CCC002S003A

	9	20			29
		12	23	16	
35	1			10	
13	36		22	27	2
21	14	7	30		34
8		33	17	28	6

3,4,5,11,15,

18,19,24,25,

26,31,32,

6CCC135S003B

			33	7	29
19	14	27			
35	12	31	17	3	13
6	25	2	24	34	20
28	23	36			
			26	30	22

1,4,5,8,9,

10,11,15,16,

18,21,32.

Solve the 8x8 MSP by filling in the numbers on the right to make the sum of each row, column, and diagonal equal to 260.

8B8004S003

4	31	37	57	3	32	38	58
45			24	46	49	11	23
	6	63	35	26	5		36
53	42	19	15	54	41	20	16
	30		60	2			59
48		9	21	47	52	10	
28		62	34	27	8		33
56				55		17	13

1,7,12,14,18,22,25,29,

39,40,43,44,50,51,61,54.

59

Week S004:

Solve the 4x4 MSP by filling in with numbers on the right to make the sum of each row, column, and diagonal equal to 34.

4AA007S004A

		8	
		13	
9	16	2	7
	11	14	

1,3,4,5,6, 10,12,15.

4AA071S004B

			9
			4
		1	15
3	13	12	6

2,5,7,8,10, 11,14,16.

Solve the 6x6 MSP by filling in with numbers on the right to make the sum of each row, columns, and diagonal equal to 111.

6BBB076S004A

31					35
18	14	1	27	32	19
29	7			25	15
22	30			12	8
9	5	10	36	23	28
2					6

3,4,11,13,16, 17,20,24,26, 33,34,35.

6BBB046S004B

2			29	18	22
34	5	30			21
24			33	1	17
20	36	4			13
16			25	32	3
15	28	8			35

0,7,9,10,11, 12,14,19,23, 26,27,31.

Solve the 8x8 MSP by filling in the numbers on the right to make the sum of each row, column, and diagonal equal to 260.

8H2009S004

		52	51	48	47	21	22
40	39	29	30			60	59
37	38	32	31			57	58
20	19	41	42	53	54		
17	18	44	43	56	55		
64	63			25	26	36	35
61	62			28	27	33	34
		49	50	45	46	24	23

1,2,3,4,5,6,7, 8,9,10,11,12,13, 14,15,16.

Week S005:

Solve the 4x4 MSP by filling in with numbers on the right to make the sum of each row, column, and diagonal equal to 34.

6AAA100S005A

	28	15	8	19	
21	14	7	30	5	
31	27	2		36	
	1			10	6
	32	25	12	23	
	9	29	22	18	

3,4,11,13,16,
17,20,24,26,
33,34,35.

6AAA003S005B

24	34			16	20
28	5				18
8	30	4	35		22
15		2	33	25	29
		36	1	32	
17	21			3	13

6,7,9,10,11,
12,14,19,23,
26,27,31.

Solve the 6x6 MSP by filling in with numbers on the right to make the sum of each row, columns, and diagonal equal to 111.

4AA064S005A

		7	14
		9	
		16	
8	13	2	11

1,3,4,5,6,
10,12,15.

4AA056S005B

9	4	15	6
			3
			13
		1	12

2,5,7,8,10,
11,14,16.

Solve the 8x8 MSP by filling in the numbers on the right to make the sum of each row, column, and diagonal equal to 260.

8C2058S005

58	1		39	57	2		40
	55	42		15	56	41	
36		6	62	35		5	61
	45	51	11		46	52	12
59	4		38	60	3		37
13	54	43		14	53	44	
33		7	63	34		8	64
	48	50	10		47	49	9

17,18,19,20,21,
22,23,24,25,26,
27,28,29,30,31,32.

16	25	35	22	48	57	3	54
20	5	63	10	52	37	31	42
62	43	49	8	30	11	17	40
1	24	14					27
47	58	4					21
51	38	32					9
29	12	18					7
34	55	45	28	2	23	13	60

Solve the 8x8 Magic Square Puzzle by placing the following numbers back into the empty grids to make the sum of each Column, row and diagonal equal to 260.

6, 15, 19, 26, 33, 36, 39, 41, 44, 46, 50, 53, 56, 59, 61, 64

There are two sets of answers.

638	693	634	693	588	589	602	597	598	233	228	229	183	183	184	197	192	193	314	309	310	269	264	265	278	273	274
631	636	639	586	590	594	595	599	603	226	230	234	181	185	189	190	194	198	307	311	315	262	266	270	271	275	279
635	637	632	591	592	587	600	601	596	231	232	227	186	187	182	195	196	191	312	313	308	267	268	263	276	277	272
575	570	571	611	606	607	647	642	643	170	165	166	206	201	202	242	237	238	251	246	247	287	282	283	323	318	319
568	572	576	604	608	612	640	644	648	163	167	171	199	203	207	235	239	243	244	248	252	280	284	288	316	320	324
573	574	569	609	610	605	645	646	641	168	169	164	204	205	200	240	241	236	249	250	245	285	286	281	321	322	317
620	615	616	629	624	625	584	579	580	215	210	211	224	219	220	179	174	175	296	291	292	305	300	301	260	255	256
613	617	621	622	626	630	577	581	585	208	212	216	217	221	225	172	176	180	289	293	297	298	302	306	253	257	261
618	619	614	627	628	623	582	583	578	213	214	209	222	223	218	177	178	173	294	295	290	303	304	299	258	259	254
66		26		22		30		396			391	350	345	348	359	354	355	719	714	715	674	669	670	681	671	679
64		72		23		28		38	392	396	343	347	351	362	356	360	712	716	720	667	671	675	676	680	684	
70		24		25		34		393		289	348	349	344	357	358	352	717	718	713	672	673	658	681	682	677	
8		4		39		80		76	327	329	366	383	364	404	399	400	658	651	652	692	687	688	728	723	724	
5		37		45		77		325		333	354	368	389	397	403	405	649	653	657	695	693	721	725	729		
6		2		43		78		74	331	325	366	367	362	402	403	396	654	655	650	690	691	686	726	727	722	
48		62		59		12		377		373	356	351	382	341	335	701	695	697	710	705	706	663	660	661		
46		54		59		10		18	374	378	379	383	387	334	358	342	694	699	702	703	707	711	655	662	666	
52		60		56		16		375		371	384	385	380	339	340	335	696	700	695	708	709	704	663	664	665	
476	471	472	431	425	427	440	435	436	557	552	553	512	507	508	521	516	517	152	147	148	107	102	103	116	111	112
469	473	477	424	428	432	433	437	441	550	554	558	509	513	514	518	522	145	149	153	100	104	108	109	113	117	
474	475	470	429	430	425	438	439	434	555	556	551	510	511	506	519	520	515	150	151	146	105	106	101	114	115	110
415	410	411	449	444	445	485	480	481	494	489	490	510	526	530	566	561	562	89	84	85	128	120	121	161	156	157
409	410	414	442	446	450	478	482	486	487	491	495	523	527	531	559	563	567	82	86	90	118	122	126	154	158	162
413	412	407	447	448	443	483	484	479	492	493	518	529	524	564	565	560	87	88	83	123	124	119	159	160	155	
468	453	454	467	462	463	422	417	418	539	534	535	548	543	544	503	498	499	134	129	130	143	138	139	98	93	94
451	455	459	460	464	468	416	419	423	532	536	540	511	545	549	496	500	504	127	131	135	136	140	144	91	95	99
455	457	452	465	466	461	420	421	416	537	538	533	546	547	542	501	502	497	132	133	128	141	142	137	96	97	92

Solve the 27x27 Magic Square Puzzle by filling the following 50 numbers back in the empty squares to make the sum of each row, column, and diagonal equal to 9855.

1	3	7	9	11	13	14	15	17	19
21	25	27	29	31	32	33	35	38	40
41	42	44	47	49	50	51	53	55	57
61	63	65	67	68	69	71	73	75	79
81	329	330	332	370	372	376	388	390	394

By: David W. K. Mei   dwkmei815@gmail.com

274	273	278	265	264	269	310	309	314	193	192	197	184	183	168	220	228	233	598	597	602	589	583	563	634	633	638
279	275	271	270	266	262	315	311	307	198	194	190	189	185	181	234	230	226	603	599	595	594	590	536	639	635	631
272		275		268		308	313	312	191	196	195	182		166	227	232	231	596	601	600	597		601	632	637	636
319		323		282		247	246	251	238			242	202	221	206	168	165	170	643	642	647		606	571	570	575
324		316		284		252	248	244	243	239	235				167	163	648	644			612	608	604	572	568	
						260	249					240	200		204	164	169	163	641		645	605		609	569	573
256		260		300		292	291	296	175		179	220		224	211	210	215	580	579	584	625	624	629	616	615	620
261		253		302		297	293	289	180		172			217	216	212	208	585	581	577			621	617	613	
254		255				290	295	294	173		177		229	222		214	213	578	583	682	623	628		614	619	618
678		683	670	669	674	715	714	719		354	359					390	395	31	30	35	22		26	67	66	71
684		676	675	671	687	720	718	712		356	352	351	347	343	396	392	388	36	32	28		23	19		68	64
677						716	717	353								393	29	34	33						70	69
724	723	728	638	687	652	652	651	655	400	389	404	364	363	368	328	327	352	76	75	80	40	39	44	4	3	8
729	725	721	693	689	685	657	653	649	405	401	397	369	365	361	333	329	325	81	77	73	45	41	37	9	5	1
722	727	726	636	691	680	650	656	654	398	403	402	362	367	360	326	331	330	74	79	78	38	43	42	2	7	6
661	660	665	706	705	710	697	696	701	337	336	341	382	381	386	373	372	377	13	12	17	58	57	62	49	48	53
666						703	702	698				387	383	378				14	10	63						46
659	664	663	704			708	695	700		335	340	380	385	384		376	375	18	15	56		60	47			51
112	111	116	103			107	148	147		517	516		508	507	512		552	557	435	440	427		431	472		476
117	113	109		104	100	153	149		522	518		513	509	505		554	550	437	433	432		424	477			468
110	115		101	106	105	146	151		515	520		506	511	510		556	555	439	438	426						474
157		161	121	120	125	85	84		562	561		528	525	523		486	484	480	435	445		443	408			413
162		154	126	122	118	90	88		567	563		531	527	523		491	487	482	478	450		442	414			406
159		159	119	124	123	83	86		560	565		524	529	526		493	492	484	483	443		447	407			411
94						143	130	129				544	543	549				417	422	483						458
95	95	91	144	140	130	135	131	127	504	500	496	549	545	540	538	532	423	419	415	458	464	460	459	455	451	
92	97	96	137	142	141	128	133	132	497	502	501	542	547	548	533	539	537	415	421	420	481	465	465	452	457	456

by: David W. K. Mei
dwkmei815@gmail.com

Solve the 27x27 Magic Square Puzzle by filling the following 166 numbers back in the empty squares to make the sum of each row, column, and diagonal equal to 9855.

11,16,20,21,24,25,27,50,52,54,55,59,61,65,72,82,87,89,93,98,102,108,114,134, 138, 139,146,160,162,156,158,180,171,174,176,178,187,199,203,205,207,209,218, 219,221,225,236,237,241,245,256,257,259,263,267,277,280,281,283,285,286,287, 288,298,299,301,303,304,306,308,317,318,320,321,322,334,336,339,342,344, 345,346,348,349,350,355,357,358,360,370,371,374,378,389,391,394,408,410,412,418,426,428,429,430,434,436,441,444,448,453,454,462,467,470,471,473,475, 479, 481,485,488,490,495,498,499,503,514,519,521,534,535,539,551,553,556,559,564,566,574,576,582,607,610,611,622,626,627,630,646,656,662,668,672,673, 675, 684, 691, 692, 694, 696, 707, 709, 711, 713

Dedicated to the Olympic in China 2008

Dedicated

to

Jean Soohoo Mei

A Beloved Wife, Mother & Grandmother

AUTHOR'S NOTE

Magic Square is no longer magic but the era of Magic Square Puzzle has just begun. This outstanding application of Magic Square will be played by millions throughout the world. I appeal to educators to adopt Magic Square Puzzles to schools' curriculums. If you are interested, email me at dwkmei815@gmail.com.

AUTHOR'S PERSONAL DATA

David Mei

Born in Canton China

B.S. degree at City College of New York in 1969

Worked for the City of New York and New York Foundation for
Seniors/ LaGuardia Senior Center

Vounteer at Jewish Community Council of Greater Coney Island/
Jay Harama Senior Center